Contents

LEFT: *The fruits of the hop.*

BELOW: *If they are supported and tied at intervals the bines can reach a height of 20 feet (6 m) or more.*

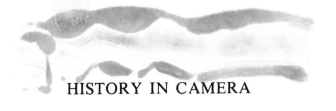

HISTORY IN CAMERA

HOPS AND HOP PICKING

Richard Filmer

Shire Publications Ltd

2

Published in 1998 by Shire Publications Ltd, Cromwell House, Church Street, Princes Risborough, Buckinghamshire HP27 9AA, UK.
Copyright © 1982 by Richard Filmer. First published 1982; reprinted 1986, 1992 and 1998. Number 6 in the History in Camera series. ISBN 0 85263 617 2.
Printed in Great Britain by CIT Printing Services, Press Buildings, Merlins Bridge, Haverfordwest, Pembrokeshire SA61 1XF.

COVER: Families picking hops. (From a nineteenth-century print by Le Blond.)

TITLE PAGE: Hop garden, Castle Hill, Farnham, Surrey.

BELOW: It is likely that the first commercial hop gardens were established in Kent. This Wealden scene of orchards, hop gardens, oasts and woodlands typifies Kent.

1. Introduction

The hop, *Humulus lupulus,* belongs to the Cannabinaceae family of plants, which also includes hemp and is related to the nettle and elm families. It is a herbaceous hardy perennial on a permanent rootstock, sometimes sending its roots down 12 feet (3.6 m) into the ground. The plant dies back to the base every year and will live for twenty years and often more. In the spring the new shoots climb not with tendrils but with tiny hairs on the stem and on the backs of the leaves. The hop twines clockwise. Given some support, the natural rigidity of the square-stemmed shoots enables them to climb to a height of over 20 feet (6 m). In the wild state, unsupported, it creeps along the ground and spirally twines itself over hedges and thickets.

The hop is dioecious, which means that the unisexual male and female flowers are borne on separate plants, and it is the larger female cones, made up of many scale-like bracts, which yield the fruits that are commercially important. As the male flower produces a good deal of pollen, for many years hop growers in England have planted one male for a hundred or more female plants in the hop garden. Pollination, however, is not necessary, for the seeds contribute little to the brewing process and seedless hops have been grown on the European mainland for many years. One advantage of this is that the brewer does not have to pay for the weight of the seed, which could be as high as fifteen per cent. Seedless hops have always been used for brewing lager, whilst brewers prefer seeded hops for English bitter. English hops are still planted with males except in Hampshire and parts of Kent, where the growers have volunteered to grow seedless plants. Hop cones bear glands which contain resins and oils; the substance known as lupulin is one of these. The resins provide the bitterness and preservative qualities and the oils delicately flavour the beer, giving it its characteristic pleasing aroma.

Beer and ale have now come to mean much the same thing, but ale, produced from malt without the addition of hops, was the original beverage of the Anglo-Saxons and English. It was not until hops were grown commercially in England in the early sixteenth century that beer, brewed with hops, began to be accepted. It became a notable British drink and the hop industry became one of the most cherished and interesting sections of British agriculture.

Ale in the traditional meaning of the word is no longer brewed. However, it is common outside Britain to refer to lager as 'beer' and to British beers as 'ale'. In this book the word 'ale' is used in its original sense and the word 'beer' as the beverage containing hops.

ABOVE: *An English ale house.*

BELOW: *One reason why Kent was suitable for early hop culture was the supply of wood available for hop poles and charcoal. The charcoal-burning industry lingered on in the hop-growing counties of south-east England until well into the twentieth century.*

2. The history of hop growing in Britain

One of the earliest recorded references to the hop was in the sixth century BC and later the great Roman naturalist Pliny (AD 23 to 79) described the hop as 'the wolf of the willow' because where hops grew among willows their twining growth proved as 'destructive as a wolf to a flock of sheep'. He also describes a delicacy prepared from young hop shoots. A botanical dictionary of 1805 contains a similar description of hops being prepared like asparagus.

A hop garden is mentioned in a deed of gift of AD 768 by Pepin, father of Charlemagne, and there are further records of cultivated monastic hop gardens in France and Germany in the seventh and ninth centuries. Many references were made to their useful medicinal properties and their value for flavouring and preserving beverages was certainly recognised in the twelfth century. Their use in beer was probably of German origin.

In the thirteenth century the land devoted to hop growing in Germany increased and by the fourteenth century hop cultivation had developed in the Netherlands, particularly in Flanders. Almost certainly hops were grown in England at this period, but only as herbs.

In the fifteenth century contemporary respect for the plant was indicated by the founding of 'The Order of the Hop' by John the Fearless and the Duke of Burgundy. John the Fearless became the twenty-seventh Count of Flanders, where he encouraged the hop planters, and to this day he is commemorated as the central character in the pageant at the annual hop festival at Poperinghe.

It is generally believed that beer first arrived in England in 1400, at Winchelsea harbour in East Sussex, in a consignment ordered by Dutch merchants working in England, who no doubt preferred their native hop-flavoured drink to the thick sweet strong English ale flavoured with herbs and spices. The Dutch influence in the trade still survives with such terms as 'kilderkin' and 'firkin' for 18 and 9 gallon (82 and 41 litre) casks.

It has been suggested that the use of hops was forbidden in England in the reign of Henry VI. In 1426 information was laid against a person at Maidstone in Kent for 'putting into beer an unwholesome weed called a hoppe'. Other towns in England tried to prevent the brewing of beer by forbidding the use of hops: Norwich did so in 1471 and Shrewsbury in 1519. Hops in drink were blamed for inciting the followers of Jack Cade's rebellion in 1450.

By the beginning of the sixteenth century beer was gaining ground. Richard Arnold's chronicle *The Customs of London,* published in 1502, contains a recipe for brewing beer with hops. Churchwarden accounts for the parish of Stratton, Cornwall, in 1514 contain the entry: 'For hoppys, the last brewing, iiijd'. Then hops were probably used not to flavour but to clarify the wort and inhibit bacterial spoilage. As the beverage lasted longer, this gave the brewer a wider market. However, hops were not grown on a commercial scale until improved varieties were introduced by the Flemings in the second decade of the sixteenth century, when the enclosure of common land made more land available for cultivation.

Tradition has it that the first English hop garden was created in the parish of Westbere near Canterbury in 1520 although there is a counter claim for a site at Little Chart, near Ashford, also in Kent. Kent was the earliest centre of hop culture for a number of reasons: the enclosed field system of farming was already established; the soils were suitable; and there was a good supply of wood for the poles and charcoal for drying. Furthermore, Kent farmers were among the most prosperous of the time and could thus afford the initial high capital outlay for establishing the gardens. An old rhyme gives an idea of the Kentish yeoman:

'A Nobleman of Cailes, a Knight of Wales
A Laird of the North Countree,
A Yeoman of Kent with his yearly rent,
Could buy them out all three.'

Because of the Kentish wool industry, many Flemish weavers had settled in the Weald and east Kent to produce the famous Kentish broadcloth and, having first-hand experience, they were able to give advice to local growers. To this day Kent still produces more hops than any other county.

By 1522 beer was brewed in England using home-grown as well as imported hops and in that year large consignments of English-brewed beer were shipped to France for the army. Army victualling accounts of 1542 show the entry '10,042 pounds of hops at 10 shillings per hundred'. Beer was also consumed on state occasions; the accounts for a royal banquet held at Windsor Park in 1528 show a provision for 15 gallons of beer at 20d and 15 gallons of ale at 2s 6d.

The old prejudices still lingered, however, and in 1530 Henry VIII forbade his brewer in Eltham to put hops or brimstone into his ale, possibly persuaded by the herbalists whose trade had decreased with the popularity of hops. The plant was also unpopular on religious grounds, for, coming from the Low Countries, it was considered by many to be a Protestant plant. 'Beer', wrote Andrew Boorde, a Sussex-born physician, in his *Dyetary* of 1542, 'is the natural drink for the Dutchman and recently it is much used in England to the detriment of the Englishman . . . it killeth those who are troubled with collic and the stone . . . it makes a man fat as shown by the Dutchmen's faces and bellies.' Nevertheless the taste for lighter, hop-flavoured, bitter-tasting beer was growing and the importance of the hop industry was recognised in 1547 when a decree stating that all arable land was to be dug up excluded land set aside for saffron and hops. Between 1549 and 1553 the government brought in experts from the Netherlands to advise English farmers on the techniques of hop growing.

Hops proved such a profitable crop that legislation was soon needed to prevent farmers from abandoning arable farming altogether in favour of hops, thus upsetting the national economy.

England exported considerable quantities of beer in the sixteenth century. The huge demand for coppice poles for the hop gardens and for the thousands of oak casks exported, never to return, prompted the first plans for timber conservation.

The first written description of English hop growing was in Thomas Tusser's entertaining manual written in rhyming couplets, *Five Hundred Points of Good Husbandry,* published in 1573.

'Where hops will grow
Here we know
Hops many will come
In a rood of room
Hops hate the land
With gravel and sand
The rotten mould
For hops is worth gold

The sun South West
For Hop Yard is best
Hop plot once found
Now dig the ground
Hops favoureth malt
Hops thrift does exalt
Of hops more reed
When time shall need.'

Around the same period Leonard Mascal, in his *Booke of the Arte and Maner of how to Plant and Graffe All Sortes of Trees,* has a chapter on hops which, anxious to praise the quality of English hops, boasts that 'One pound of our hoppes dried and ordered will go as far as two poundes of the best hoppe that cometh from beyond the seas.'

Reynolde Scot

The first English book completely devoted to hop growing was written by a remarkable man, Reynolde Scot. Born in 1538 and educated at Oxford University, he inherited lands at Smeeth and Brabourne in east Kent and cultivated them in a most enlightened

In Reynolde Scot's day the hop sets were planted in April but today planting and replacement usually takes place in the autumn.

way. At the age of thirty-six he wrote a useful and important treatise, *A Perfite Platform of a Hoppe Garden,* which made an outstanding contribution to the English hop industry, helping to establish and extend hop growing in Kent, and which gives a detailed insight into Elizabethan hop growing. Three editions appeared, in 1574, 1576 and 1578, with fine woodcut illustrations and this small book was to remain the standard reference work on hops for many generations.

Scot described and illustrated every stage of cultivation with complete confidence: 'There is no reason why hops cannot be grown on English soyle . . . I see the Flemings envie our practice here-in, seeking to cramme us with the wares and fruits of their countrie . . . and doe anye-thing that myght putt impediment to this purpose, dazeling us with the discommendation of our soyle, obscuring and falsifying the order of this mysterie, sending us to Flanders as farre as Poppering for that which we may find at home on our owne backsides'.

Much of his advice still holds good today: 'Lay the ground level, square and uniform. If grassy, rough or stiff, so first plant with hemp or beans . . . till it with plough or spade in early Winter the year before planting and every year after . . . assuring yourself that the more paynes you take, and the more cost you bestowe rightly hereupon, the more that you doe double your profite, and the nearer you resemble the trade of the Flemming.'

He calculated the likely yield and profit per acre: 'To be resolved

on all these poyntes you must make your account in this wyse . . .
One man may well keepe 2,000 hilles and yet reserve his Wynters
labor for any other purpose . . . Upon every acre you may erect
seaven, eyght or nyne hundreth hylles . . . Upon every hyll well
ordered, you shall have three pounds of hoppes at the least' (a good
crop, even today) . . . Two pounds and a half of these hops will largely
serve for the brewying of one quarter of mault . . . One hundred
poundes of these hoppes are commonly worth 26s 8d (about £1.33
per hundredweight) so as one acre of grounde and the thirde part of a
man's labor, with small costs besides shall yeelde unto him that
ordereth the same well fourty markes yearly (about £37) and that for
ever . . .'

Hop sets at this time cost about 6d per hundred. Two or three roots
were planted in a hole about 1 foot (300 mm) square and 1 foot deep,
filled in with fine mould. The holes were 8 or 9 feet (2.4 to 2.7 m)
apart, a planting distance still common today, for it allows the hops
plenty of light and air. Then sets were planted in April but in later
years they were planted in October or November.

Scot advised the erection of poles, as soon as the hops appear
above ground, at the rate of three or four to the *hill*. Straight alder
poles that have been felled in winter, when the sap is down, were used,

Scot's practice of tying the hops to the poles by binding them with rush continued until recent times.

Of Tying of Hoppes to the Poales.

ABOVE: *'Cut them assunder with a sharp hooke and with a forked staffe take them from the poles . . . You may make the forke and hooke one apt instrument.' Reynolde Scot, 1574.*
BELOW: *'The forke and hooke' described by Scot has, in the form of the binman's hook, probably been in continuous use since his day and is still occasionally used in modern hop gardens for bringing down hops on the outside alleys, where it is difficult for the tractor to gain access because of the straining wires.*

cut between 'Allhallowentyde and Christmas'. Make sure that 'your poales be streight without scrags or knobbles. Hops seem more willinglye and naturalle to encline to alders than any other kind of pole, costing not much more then fueling and with care can last six to seven years'. They should be 9 or 10 inches (250 to 280 mm) round at the base and not more than 15 or 16 feet (4.6 to 4.9 m) high unless grown in very rich soil since 'the hoppe never stocketh kindlye untill it reache higher than the poale and return from it a good yarde or too'.

The poles were placed 2 or 3 inches (50 to 75 mm) from the principal root and were sunk 1 foot to 1 foot 6 inches (300 to 450 mm) into the ground, set leaning outwards a little. A yearly supply of two loads, each containing 150 poles should maintain 1 acre (0.4 ha) continually, whilst old broken poles serve as a fuel for the oast.

Scot describes tying the hops to the poles by binding them with rush or grass that has been well toughened in the sun and then training the hops round the poles, 'directing them alwayes according to the course of the sunne'.

From then until the hops were picked, work would be continuous in keeping down weeds and raising the hill around each bine, eventually to a height of about 3 feet (900 mm). This was done with a tool rather like a long-handled cooper's adze or a mattock which would pare away the ground in the alleys and bank it up on the hills, thus accomplishing weeding and hill raising in one operation. Today the ground immediately around the hop is still known as a hill, but bears little resemblance to the huge flat-topped 3-foot mound of Scot's day.

Scot advised that at harvest time the bines should be cut close to the tops of the hills and stripped from the standing pole. 'Cut them assunder with a sharpe hooke and with a forked staffe take them from the poles . . . You may make the forks and hooke one apt instrument to serve both these turns as is here shewne.' The tool illustrated in the woodcut has changed little over the centuries and is very similar to the tools still in use in at least one hop garden in east Kent.

The hops were gathered into blankets or baskets to be carried to the oasts. Possibly the blankets were the forerunners of the hessian bins used in the Weald, west Kent, Hereford and Worcester, and the surplices of Surrey and Hampshire. Baskets were always used in east Kent until the end of hand hop picking in the 1970s.

Scot also warned against allowing animals to stray into the hop garden: 'You may arm every hil with a fewe thornes to defende them from the annoyance of Poultrie, which many times will scrape and bathe amoung the hilles . . . but a goose is the most noysome vermine that can enter into the garden, for a goose will knibble upon every young hoppe budde . . . which will never grow afterwardes and

therefore as well to avoyde the goose as other noysome cattel . . . let
your closure be strong and kept tight.'

Reynolde Scot's book written in the charming and vigorous
Elizabethan style affords a remarkable testimony to the knowledge of
hop culture of the period.

Seventeenth century

Towards the end of the sixteenth century, ale was described as
'thick and fulsome, and no longer popular except with a few'; beer
was established. Hop growing had spread rapidly; farmers
throughout England were incorporating small acreages of hops on
their farms, and a deed of transfer as early as 1592, relating to land in
Caernarvonshire, mentions hops, showing that hop cultivation had
already spread into Wales.

An Act of Parliament in 1603 was concerned with 'Frauds and
Deceipts practiced by Forreiners', who evidently included 'leaves,
stalkes, straw, leggets of wood, dross and other soils' with their hops.
To prevent 'similar uncleaned hops packed in this country' the
merchants buying them were ordered to forfeit their value. This Act
could well have maintained the quality for in 1629 it was said that
English hops were the best in the world.

In *The Herball or General Historie of Plantes,* published in 1633,
the manured or garden hop and the wild or hedge hop are both mentioned. 'The manifold vertues of hops do manifestly argue the
wholesomeness of beers above ale; for the hops rather make it a
physicall drink to keep the body in health, than an ordinary drinke for
the quenching of our thirst.' Even so, one writer described beer as a
'Dutch boorish liquor'.

The appearance of the hop grounds changed during the seventeenth
century when the hills were reduced in size by as much as half. They
were manured with earth from dunghills, soap ashes and bracken, all
of which were rich in potash — an important nutrient for hops. One
writer suggested mixing ox blood and lime in the hills, which would
'give comfort and encouragement to the plant and save the roots from
worms and other vermin'.

Most manual work such as the dressing of the hills, hoeing, poling
and tying was carried out by contract at an agreed price, often about
40 shillings per acre, but the contract excluded pulling, picking, drying and bagging, which were charged by the day. An acre (0.4 ha) of
good hops could bear 11 to 12 hundredweight (550 to 600 kg), which
could fetch as much as £40 or even £60. The dried hops were packed
in $2\frac{1}{4}$ hundredweight (114 kg) bags and sold at local fairs, Sturbridge
in Cambridgeshire being the best known.

LEFT: *Spraying in a Hampshire hop garden. Even in the seventeenth century attempts were made at pest control and today some gardens are sprayed as often as ten times a year against mildew, mould, aphis and other pests.*

RIGHT: *The hop market in the Borough. Towards the end of the seventeenth century a hop market was established in London, first at Little Eastcheap, then in the Borough.*

Once a hop garden was established, it might well be let to a hop master. By 1655 hops were grown in fourteen counties of England (although one third of the crop was produced in Kent) but there were still not enough to satisfy the demand and a great quantity of Flemish hops was imported. Some farmers would not grow hops because of the erratic yields caused by drought, wet periods and mildew. An old Kentish verse on hops includes the lines:

'First the flea, then the fly,
Then the mould, then they die.'

Yet in a successful year an acre of good hops could be more profitable than 50 acres of arable land.

Towards the end of the seventeenth century beer began to be bottled in quantity. One of the earliest brewers to do so was George Cross from Woburn, who bottled for the Duke of Bedford. Writers spoke of different varieties of hops but they were not named. Some attempts were made at pest control and often the plants were sprinkled with water in which wormwood had been boiled.

A hop market was established in London at Little Eastcheap, later to be transferred to the Borough in south-east London, which was more convenient to the Kentish growers.

Eighteenth century

In 1710 duty was imposed on hops for the first time at the rate of 1d per pound on English hops and 3d per pound on Flemish hops. The Act also prohibited the use of any bittering ingredient other than hops in brewing beer intended for sale, as hops were far more wholesome. The duty was successful in yielding a large revenue. The actual duty imposed varied from year to year and thus appealed to the eighteenth-century gambling instinct and for the whole of the century speculation on the hop tax was a popular form of betting. Import duty encouraged smuggling and from 1734 the penalty for illegal export was not only the destruction of the hops but a punitive fine of 5 shillings per pound on the hops confiscated. Annual yields varied from 2 to 15 hundredweight (100 to 750 kg) per acre, according to weather, pests and diseases, and this contributed to the great variation in revenue from hop duty.

Daniel Defoe's description in 1724 of the extensive hop plantations in Herefordshire appears to be one of the first references to the Midland hop industry. By 1781 the rectory at Martley in Worcestershire was said to be one of the most valuable holdings in England, where the profits from hops could amount to £1,000. Hops are not grown in Martley today although they have been grown nearby in recent years at Suckley, Alfrick and Leigh Sinton. Defoe also noted the huge hop gardens around Maidstone and at Canterbury, where there are still hop gardens within a mile of the cathedral and oasts within the old city wall. In the Canterbury district there was a suburban development of very small hop gardens cultivated by the tradespeople themselves: the saddler, the grocer and even the housewife often had a hop garden of maybe only half an acre (0.2 ha) or so.

In Defoe's day Sturbridge Fair in Cambridgeshire was the principal fair in England specialising in the sale of hops and wool, although it was later to be superseded by Weyhill in Hampshire. The price of hops was apparently governed by the price achieved at Sturbridge, which, although far away from the principal hop-growing counties, was readily accessible by water both from the north and the south. Kentish growers obviously used the London hop market. By 1765

Hop picking from an eighteenth-century print. 'Plenty of hands should, on all accounts, be provided for this important business: women do it as well as men: it is a work rather of care than of labour' (The Farmer's Kalendar, Arthur Young, 1771).

hops were sold at twenty-five fairs in England and Wales.

Kent was still producing more hops than any other county and there fruit trees were sometimes planted in the alleys between the hops. When the trees began to mature the hops were removed. This practice continued until the twentieth century. By 1750 paler beer was coming into vogue and as the lighter-coloured hops from the Farnham area in Surrey were favoured by brewers many Kentish farmers replaced their Canterbury hops with the Farnham varieties. By 1763 it was considered that $1\frac{1}{2}$ pounds of Kent hops were equal to 2 pounds of Worcester hops. Kentish hops were used in beer brewed for keeping and Worcester hops in beer to be drunk within a month to six weeks.

Nathanial Kent in 1775 wrote that a successful crop, though profitable, required a great deal of attention and a large quantity of manure. The farmers of Hereford and Worcester applied a great part of their farm manure to the hop yards, at the expense of the remaining land, which Kent considered most unwise because the rest of the farm brought in certain profits in comparison with the uncertain profits of hops. It was said to cost around £15 an acre to cultivate hops and although in some years there were returns of £50, £60, £80 and even £100 per acre this extraordinary profit was very uncertain and there were often many years of failure. In 1780 many hops were grubbed around Canterbury because of low prices.

Hops were probably grown in the west midlands from the early seventeenth century and when Daniel Defoe visited Herefordshire in 1724 he wrote of extensive hop plantations. At this Herefordshire farm the traditional oasts are in the centre of the photograph and a modern oast to the left, with a louvred vent along the ridge.

When hop duty was high and hops were scarce, brewers began to experiment again with herbal preservatives such as willow bark, gentian and walnut leaf but tastes had changed and none were acceptable or efficient alternatives.

To prevent frauds in marketing, an Act was passed in 1774 which required the bags or pockets in which the hops were packed to be marked with the year, place of growth and the grower's name. Growers marked their own pockets with stamps and stencils and many of them devised their own symbols. The Farnham growers became well known not only for the quality of their hops but also for their packaging.

The late eighteenth-century revival of British agriculture focused attention on hop varieties. Popular at the time were Williams hops from Farnham in Surrey and Jones hops, introduced in about 1780. At this period the famous Golding was discovered and the Farnham and Canterbury White Bines displaced older varieties.

Hops were manured every third year with woollen rags brought in from London, costing about £5 a ton and spread at a rate of about 1 ton per acre.

ABOVE LEFT: *The Surrey growers from the Farnham area were renowned for the fine presentation of their pockets. This pocket stamp is typical of the Surrey and Hampshire hop-growing districts.*

ABOVE RIGHT: *A Kentish pocket stencil.*

BELOW: *Today the main marking of pockets is carried out by the supplier.*

Consumption of hops during the century was considerable and in some years a million barrels of beer were brewed in London and the suburbs alone. 4 pounds (1.7 kg) of hops were used per barrel — more than twice the present use.

By the end of the century hops were grown as far north as Aberdeen but a writer of the day said that brewers would prefer to purchase their hops from England, which they felt was better suited to raising hops of quality. However, it was a Scottish gentleman farmer, H. Hume, who in 1776 advocated a system of hop training which would reduce the huge expense of poling. He suggested preparing a line of poles running east to west, 8 or 9 feet (2.4 to 2.7 m) high, and, instead of allowing the hops to climb the poles, training them west to east to run at angles between the poles. There is little evidence that Hume's system became popular and it was at least another hundred years before a satisfactory system of wire and string work began to oust the old poling method.

Nineteenth century

The nineteenth century was the golden age of the hop industry. In 1800 there were 35,000 acres (14,000 ha) of hops in Great Britain and by 1850 50,000 acres (20,000 ha). In 1862 the hop excise duty, which had been between $1\frac{1}{2}$d and 2d per pound, was removed, and by 1870 hops were cultivated in forty English, eight Welsh and five Scottish counties, although most of the acreage was in Kent. Hop acreage reached its peak in 1878 with 71,789 acres (29,053 ha), after which, apart from some expansion in the Midlands, the acreage decreased, and by 1900 it was reduced to about 50,000 acres (20,000 ha).

Early in the century Mr Colgate developed the Colgate hop from a specimen found growing wild in a farm hedge at Chevening in Kent. Other popular varieties were Golden Tips, Canterbury Grape and Mayfield Grape. Later Jones, Cooper's White and Amos Early Bird were grown in some quantity together with the famous Fuggle, a mid-season hop of good size and quality extensively grown in the Weald and the Midlands, though its acreage is now rapidly dwindling. It was said to have originated from a seed thrown out with some crumbs from a hop picker's dinner basket on the farm of George Stace of Horsmonden in Kent in 1861 and was subsequently introduced as a commercial variety by Richard Fuggle in 1875.

Profits in the industry were still erratic and fortunes were made and lost. Jeremiah Smith, seven times Mayor of Rye in East Sussex, was one of the largest hop growers in England and it is said that from 1850 until the repeal of hop duty in 1862 he paid £33,462 in hop duty. For over twenty years he farmed 1,200 acres (485 ha) of land in

Many of the hop gardens that remained on the pole work system towards the end of the nineteenth century evolved a system of stringing without the use of overhead wires. This practice continued into the 1970s. The stringer here is throwing a 'lobber' over the top string.

Mould was a frequent problem and about 1850 it was discovered that dusting the hops with sulphur helped to control the disease.

Kent and Sussex at an estimated rental of £4,000. The decline in the value of hops contributed to his bankruptcy when he died in 1864. In 1827 the cost of a new hop garden was between £80 and £90 per acre for the first year, then £30 to £40 per annum. Rows were planted 8 feet (2.4 m) apart with 6 feet (1.8 m) between the hills and three or more plants per hill. In the Farnham district it was common practice to prune in midsummer and the cut bines were used to feed cows.

Further attempts were made to discover a remedy for flea beetle and aphid spread but none was effective. A remedy for wireworm was to bury half a potato each day to tempt the wireworms away from the tender hops. About 1850 it was discovered that dusting with sulphur helped control mould. This was an unpleasant task and the thick, sticky stains were removed from the operators' hands with crushed hop leaves. A clumsy horse-drawn powdering machine was subsequently invented which propelled the sulphur dust over the hops. In 1848 ladybirds were introduced to keep down hop flies. Another remedy was to knock the flies into a tar-covered tray. From about 1865 soft soap solution, tobacco juice and later quassia, applied from a hand-pumped spraying machine, were used as insecticides. After 1883 hop spraying became general.

Newly planted hop gardens were often ploughed to a depth of 16 inches (400 mm) with a strong plough drawn by perhaps ten horses. After this only the horse *nidget,* a small triangular cultivator with a number of harrow-type prongs, drawn by one or two horses, was allowed in. The nidget could not get sufficiently near to the plants, so the areas around them were hoed with a small short-handled hop hoe and dug over by hand with the *spud,* a heavy three-pronged fork. Digging, hoeing and nidgeting often had to be done as many as eight times a season.

The most important development of the nineteenth century was in the system of training the hops. Chestnut was generally used for hop poles, which, in the middle of the century, cost about £1 per hundred, about four thousand being needed per acre. They had a limited life as creosote was not generally available until about 1862 and even then there were many objections to creosoting the lower parts of the poles. Estates and plantations of regularly coppiced chestnut were maintain-

RIGHT: *In the latter part of the nineteenth century hops were sprayed with various solutions, including soft soap and quassia. The solution was applied either from a hand-pumped spraying machine or from a horse-drawn hop washer like the one shown here.*

BELOW: *A wooden-framed horse-drawn nidget.*

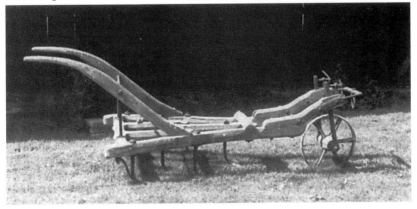

Taking the horse hoe out into a west midland hop yard. Note the 'cobweb' effect of the strings tied across the poles to give some rigidity.

ed for a regular supply and these coppices in the south-east of England provide many of the larger chestnut poles used for wirework today.

Training systems were tried, including the American method of stringing, whereby short poles were erected with strings running along the top of the pole. This method was not successful because it was against the nature of the leading stem to climb of its own accord and frequent tying was necessary.

In 1874 Coley introduced his vinery system, which had some success. Two poles about 8 feet (2.4 m) high were placed in the ground either side of each hill; a crosspiece joined the uprights a third of the way up, and two pieces of thick wire lashed to their tips formed a fork. Diagonal poles were placed in the socket of one pole and rested in the wire fork of the opposite pole. The system worked, but as it cost almost £90 per acre to equip a garden it was not popular.

The simplest and most economic method was the wirework system introduced by Henry Butcher of Sheldwich near Faversham in Kent in about 1875, and it became very popular in east Kent. Although the poles are higher, the overall pole density is less. Stout horizontal wires were attached to chestnut poles near ground level, at breast level and at the top of the poles. The hop bines could spiral up coir strings

which were taken from the bottom wire to the breast wire, then sloped across the alley to the top of the neighbouring row. To speed the tying of the strings to the top wires, a horse-drawn trolley with an elevated platform was used or, especially in Kent, the stringers walked on high stilts. Henry Butcher's method quickly gained popularity and the system spread to all hop areas and in 1884 cost about £50 an acre to install. The system is still used by one or two growers in east Kent,

A nineteenth-century string system.

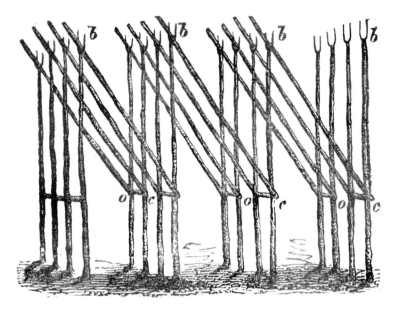

ABOVE: *Coley's vinery system, introduced in 1874.*
BELOW: *The Butcher system was invented by Henry Butcher in 1875. The stilt walker ties the strings to the top wire whilst at ground level the strings are caught with a stick and tied across to the breast wire on the other side of the alley.*

Although umbrella and Worcester work has generally superseded the Butcher system, the method is still used in one or two hop gardens.

slightly modified to suit machine picking. The old-fashioned pole system, however, lingered in some areas until the 1970s.

By the end of the century Wye College, founded in 1447, had become an agricultural college and it included hop culture in its studies.

Twentieth century

By 1909 the area of land growing hops had fallen to 32,000 acres (13,000 ha), less than half that of twenty-five years earlier. During the First World War brewing was considerably reduced, and to avoid a huge surplus the government further restricted the amount of land under hops and introduced a system of hop control, which remained in force until 1925. To provide some protection to the industry, a customs duty of £4 per hundredweight was imposed on foreign hops.

The period from the late 1920s to the early 1930s was disastrous for hop growers, who had to contend with the depression, surplus hops, low prices and two difficult diseases — hop downy mildew and verticillium wilt.

In 1932 the Hops Marketing Board was created. The producer-controlled board consisted of fourteen members elected annually by

ABOVE: *At the beginning of the twentieth century the English hop industry was suffering badly because of the free importation of foreign hops. In 1908 huge demonstrations took place in London and Kent in favour of a duty on imported hops of £2 per hundredweight (50 kg).*

LEFT: *A commemorative paper handkerchief was produced for those who attended the Trafalgar Square demonstration.*

the hop-growing districts of Kent, Hampshire, Herefordshire, Sussex and Worcestershire, four special members elected annually by hop producers and two nominees of the Minister of Agriculture. The board exercised a monopoly control and was immune from the Restrictive Trade Practices Act, thus ensuring a sheltered market for its producers. In 1982, following new legislation to conform with EEC rules, the Hops Marketing Board Limited, a voluntary agricultural co-operative, took over all assets and procedures of the old board. In 1985 the Board was restructured and became English Hops Limited.

The English hop-picking machine was made in 1934 but did not begin to threaten hand picking until after the Second World War,

Shimming the alleys with a horse hoe, about 1904. Note the pole system of at least two poles to each hill.

when it was generally accepted in the Midlands. It is now used in most hop gardens as it allows picking to continue in all weather conditions. The bines are cut about 2 feet (600 mm) above the ground. A crow's nest attached to a tractor-drawn trailer carries a cutter which slashes the strings from the overhead wires. The bine and string fall into the trailer below, which is then driven to a covered picking plant. The bines are hung on a continuously moving overhead track taking them into the picking machine, which plucks the hops from the bine and sends them through a system of conveyor belts and rollers. Eventually the hops are emptied into the poke and thence into the oast or drying unit.

Permanent wirework systems, such as the Worcester or the umbrella systems, where the strings are banded in out of the way of the tractors, have become established. The stilt-walker may carry out maintenance but not stringing. The more modern systems have S hooks clamped on to the top wires and corkscrew galvanised wire pegs with a pig's tail curl at the top twisted in the ground near the hop plant. A continuous length of coir string is wound into 8 pound (3.6 kg) balls, which are carried by the stringer in a canvas pouch round his shoulders. Carrying the long rod or cane with a piece of bent piping fixed to the end, the stringer deftly threads the string through his stringing rod over the S hooks on the top wire and gently sweeps the string down to ground level, catching it in the twist of wire embedded in the ground. Each hill requires its own stringing system and with 1,200 hills to the acre some 15 miles (24 km) of string are used. Hop stringing is still a totally unmechanised rural craft requiring infinite patience, the more impressive because it is only practised for a few weeks of the year.

Although experiments have been carried out with polypropylene string, coir is still generally used because it stretches a little when wet, thus taking more strain when it is most needed, and shrinks in drying, tightening up the whole system. Its hairy surface is perfect for the hairs on the hop bines and leaves to grip.

Hop dressing, to cut back the underground runners of the hop set early in the year to contain the rootstock, has now been discontinued. Each hill was dressed individually by pulling the earth away from the crown with a three-tined Canterbury hoe and cutting round the hill first with a sharp-bladed knife and then with a tommy hoe. Finally, the earth was drawn back round the rootstock. Nowadays, with the use of chemical sprays to control weeds, ploughing and cultivation have been eliminated from established hop gardens.

In 1947 a separate department of hop research was established at Wye College, which is now acknowledged as a world leader in most

Hops trained on a modern Worcester system.

TOP: *'Banding in' umbrella work. The strings are pulled together at a height of about 5 feet (1.5 m) so tractors have clear access up the alleys.*
ABOVE: *Galvanised screw pegs are twisted into the ground and the coir string is looped through the twist at the top.*
RIGHT: *Stringing with a stringing rod.*

RIGHT: *Modern wirework systems are designed on engineering principles although some of the methods and techniques used are traditional and practical. A long-forked stick guides the wire over the end posts. Note the home-made wire dolly in the foreground.*

LEFT: *Hop stringing on stilts is one of the most impressive rural crafts, especially as it is only practised once a year. It is said that this stilt walker could string an alley of hops in much the same time as one could walk over the rough earth at ground level.*

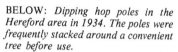

BELOW: *Dipping hop poles in the Hereford area in 1934. The poles were frequently stacked around a convenient tree before use.*

ABOVE: *Selected straight-growing chestnut poles for wirework laid out in a hop garden to await bark-stripping. A pole of this diameter will probably be used for 'outside bat'.*
BELOW LEFT: *Winter work in a pole garden. The pole holes are made with a 'pitcher' or 'peeler'.*
BELOW RIGHT: *Hop dressing. The hop set is exposed by raking away the soil with a two- or three-tined hoe. The bases of the bines are cut away with a sharp dressing knife, together with unnecessary rootstock.*

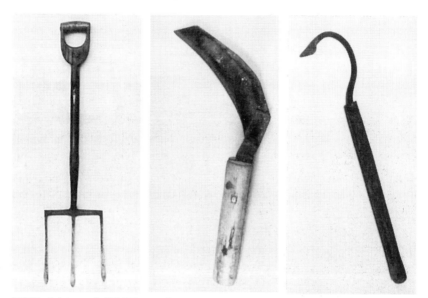

LEFT: *A hop spud.* CENTRE: *A home-made hop dressing knife.* RIGHT: *A tommy hoe.*

aspects of hop research. It has its own hop garden and oasts. As hop yields are so variable long term planning is difficult. Among other projects, scientists are investigating overhead sprinkling, piped irrigation and permanent watering systems to assist more uniform yields.

Methods of propagation are changing. Whereas it used to take at least two years before plants from base cuttings were established, now with modern propagation methods, it takes only one year with softwood cuttings from shoots grown in the same year. The small cuttings are placed in compost under controlled heated and misted conditions – 22 C (72 F) is the optimum temperature – until they take root, usually in eight to ten days, then they are hardened off. If planted out in April or May, bines may reach the top of the strings by the end of the first season and bear hops the first year.

The hop is a 'short day plant' which changes from the vegetative to the fruiting phase, as the days get shorter. With tungsten lighting to provide sixteen hours of daylight each day, it is possible to propagate continuously and so experiments can be carried out all year round.

Wye College produces thirty thousand seedlings a year, which are eventually reduced to about two hundred, and a new variety is

ABOVE: *The Hop Research Department at Wye College in Kent has been carrying out experiments for the hop industry since 1894 and produces some thirty thousand seedlings a year. Under controlled heated and misted conditions it is now possible to establish a plantation in one year with softwood cuttings.*

LEFT: *The bines are hung on to a continuously moving overhead track that takes them to the picking chamber.*

probably chosen once in five years, so 150,000 seedlings are often grown to produce one new variety. Because of the danger of wilt, some propagation takes place in areas such as East Anglia where hops are not normally grown.

Although the area of land growing hops has dropped substantially, with new varieties and methods the crop has dramatically increased, thereby releasing land for other purposes. New varieties have been developed not only for improved qualities of bittering and preserving but also for wilt resistance and wilt tolerance: hop growing in the Weald of Kent, for example, only survives with the wilt resistant varieties. The old favourite, the Fuggle, almost died out, although the brewers liked it because it was a true and tested variety. In the 1930s eighty per cent of the hop acreage produced Fuggles; today this has dropped to ten per cent and continues to fall due to the spread of wilt into the West Midlands. It owes its survival in part to the Campaign for Real Ale (CAMRA). Brewers are using more high alpha hops, that is those with improved bittering quality provided by new

A hop drying installation in a Herefordshire hop farm, 1980.

varieties such as Wye Target, rather than the old traditional ones.

Much progress has been made in the design of drying kilns. Hop growers have invested in sophisticated equipment such as three-tier batch dryers or continuous dryers, some drying plants consisting of a chamber about 47 feet 6 inches by 9 feet (14.4 m by 2.7 m), where each tier can be loaded with hops 18 inches (450 mm) deep so that when the dryer is fully loaded the equivalent of 4 feet 6 inches (1400 mm) of green hops will be drying at a time instead of the normal 18 inches to 2 feet (450 to 600 mm) in the traditional kiln.

Processing hops into extracts or pellets is another important development. Hop extracts were first manufactured in the United Kingdom in the First World War, and since the 1980s, because of the high cost of storage, handling and transport, only about one third of the hops produced are used in their natural state.

To economise on the high cost of setting up a hop garden and harvesting, some new varieties are being bred to a height of only 8 feet (2.4 m). In recent years the traditional hop varieties are regaining favour, and some other growers are beginning to operate a system of 10 foot (3.0 m) low trelliswork, which has economic advantages whilst still using traditional hop varieties.

BELOW LEFT: *Plenty of jobs have to be carried out by hand. Women with nimble fingers carry out the second training when the hops are about 5 feet (1.5 m) high.*

BELOW RIGHT: *Machine harvesting on 10 foot (3.0 m) low trelliswork on an East Sussex hop garden in 1997.*

3. Hop picking

Kentish baptismal records of the mid seventeenth century mention some 'strangers who came a hopping,' indicating that there was not enough local labour to harvest the crop. Soon pickers were travelling from the Black Country, Wales and even Ireland to the west midlands and from all parts of London and beyond to the hop gardens of Kent, Sussex and Hampshire. With the coming of the railways, hoppers gathered at stations waiting for the 'hop pickers' special.' Whole families came, hundreds of men, women, children, babies and dogs from slum areas, with their few possessions slung over their shoulders. Local villagers were wary of them and tried to protect their children from the ribaldry and blasphemy of the London pickers. The Londoners, however, were reliable and there are many records of families visiting the same gardens for several generations. Frequently grandmother, daughter and grandchild were to be found working around the same canvas bin or baskets which were moved forward as the pole-pullers cleared the bines. Children unable to reach the top of the bin picked their hops into an open umbrella.

In the days of the old pole system, the poles themselves were uprooted with the bines still clinging and laid across the bin or basket. The cones were snatched off and dropped into the bin. With later systems of stringing the picker or the binman would pull down the hop bine with the binman's hook and the pickers would lay the bines across their laps and pluck each hop cone individually with the thumb and middle finger at great speed. Often an experienced picker, who had a bine that was comparatively free of leaves, developed a technique of stripping the length of bine, with the forefinger and thumb, in one action.

Although there are many reports of the jolly parties of hop pickers throughout the industry, living conditions were extremely poor. Frequently the only shelter the pickers had was canvas slung across sheep hurdles although some farmers provided proper tents and William Elys in *Modern Husbandry* (1750) refers to a Kentish grower who provided huts furnished with wheat straw bedding. The picking rate in those days was 1d a bushel. The accounts of Sir Edward Filmer's hop garden at East Sutton in Kent in 1739 show that hops were sold at 50 shillings per hundredweight (50 kg). A century later, in 1838, the rate had risen to about $1\frac{1}{2}$d a bushel and it was said that a family of five could earn between 7 and 10 shillings a day. The Reverend J. J. Kendon, visiting Goudhurst, a village in the Weald of Kent in the centre of the hop country, was appalled at the plight of the

ABOVE: *Hop pickers arrive at the railway station.*
BELOW: *Many pickers remained faithful to one particular hop garden and there are records of pickers visiting the same gardens for seventy years or more without a break.*

Three generations of the same family in one of the last hop gardens to carry on the tradition of hand picking.

pickers and began to campaign for improvements. In his first report in the 1860s he wrote: 'They sleep ... almost like the cattle in the field. To mingle with these poor creatures, to see their habits and hear their language; to witness the awful length to which they go makes it seem almost impossible that we can be living ... in the nineteenth century.'

Mr Kendon made his headquarters at Curtisden Green near Goudhurst and by 1889 had a team of over a dozen missionaries. Later other charitable and religious missions took up the work and through the efforts of such bodies as the Society for the Conveyance and Improved Lodging of Hop Pickers and the Church of England Temperance Society Mission to Hop Pickers, and by local bylaws, specially built hopper houses began to be constructed which gave each person a minimum space of 16 square feet (1.5 sq m) to live in.

ABOVE: *The pickers were organised into orderly rows known as 'drifts'.*

BELOW: *The pole laid across the top of the bin was known as the 'horse pole'. Upon it the pole with the bine attached could be rested for picking.*

ABOVE LEFT: *The pole puller with a hop dog. The bines were cut at the base and the poles were prised up with the serrated teeth of the hop dog with the bines still clinging.*
ABOVE RIGHT: *A hop knife for cutting the bines.*
BELOW: *A hop dog.*

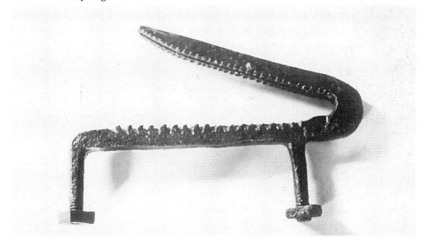

A tallyman and a group of binmen with binmen's hooks. With later systems of stringing and wirework, the strings were cut at the top with the sharp blade of the binman's hook. It was a matter of pride to learn the particular jerk most likely to enable the bine to come down clean.

In 1898 Father Richard Wilson, an Augustinian priest from Stepney, became curious about what his parishioners did when they disappeared in September and persuaded one of the families to take him with them on their annual visit.

He, too, was horrified but by continuing to work with the pickers and live in similar conditions he was able to gain their confidence. After several years he acquired a redundant public house at Five Oak Green in Kent, which he converted into a Hoppers' Hospital. For nearly sixty seasons the hospital provided treatment and the building stands today as a monument to the man and a reminder of the need it filled.

Traditionally pickers were paid by the shilling tally, so many bushels to the shilling, the lower the number, the better the pay. The tally was fixed according to the ease or difficulty of picking from year to year and differed from farm to farm. Before 1914 the usual tally was seven or eight bushels to be picked for one shilling; after the war it became four or five bushels, but by 1930 the old seven or eight bushels were again required for a shilling pay. It is difficult to state individual earnings for it was not often that the picker worked alone at a bin. In 1891 there is a record of a woman picker earning £2 19s $2\frac{1}{2}$d

ABOVE: *As early as 1883 the Church of England Temperance Society Mission to Hop Pickers had started temperance work in the hop gardens.*
BELOW: *Father Richard Wilson was possibly the first London East End priest to go hopping. He was horrified at the squalid conditions and particularly at the lack of medical facilities. In 1910 the 'Hoppers' Parson', as the newspapers called him, rented the Rose and Crown, a redundant public house. Later he managed to buy the property and the building became known as the Hoppers' Hospital. It still stands at Five Oak Green in Kent.*

ABOVE: *It was through the efforts of such bodies as the Society for the Conveyance and Improved Lodging of Hop Pickers that specially built hopper houses began to be constructed which gave each person a minimum of 16 square feet (1.5 sq m) to live in.*

LEFT: *Although hop picking is not hard physical work, thousands of hops must be picked to fill a bin or basket. This entails a lot of standing, especially around the bins, where the pickers might have to stand for ten hours a day.*

in three weeks and in the same year a pensioner, his wife, son and daughter, picking at one bin, earned a total of £4 12s 2d in four weeks.

In 1931 George Orwell spent a week hop picking at a farm near Maidstone in Kent and wrote that he was so cold in his hut that he had to steal some pokes (hop sacks) to keep him and his companions warm. Evidently he managed to earn 9 shillings in a week and recalls that a family of gypsies working nearby, who had been picking every season since birth, earned 14 shillings each.

A family of two adults and three children in the late 1930s earned about £2 10s to £3 per week and many families might take home £10 to £15 at the end of the season, as living expenses were low, with shelter and fuel provided. One of the conditions of employment was that the hoppers must stay the full harvest term. To ensure this some pay was kept in hand each week and paid in a lump sum when the work was through. Many pickers returned to the same farm year after year, booking ahead from one year to the next.

Between the wars conditions improved immensely. There were mission huts of many denominations in most of the larger hop gardens and the Red Cross had camps in the fields from 1923. In 1936 at Whitbreads' largest hop garden at Beltring in Kent conditions were described as 'ideal'. The huts provided hot and cold water; there were excellent sanitary arrangements and a surgery at hand. This particular garden became a favourite of the London pickers, although they had to behave. Drunkenness, swearing and rough acts were reported to the manager, who kept a 'black book' and thus managed to control five thousand pickers every year, who picked half a million bushels of hops from half a million hop hills on over 300 acres (120 ha) of land.

The Londoners worked extremely hard in all weathers (no picking, no pay) from dawn to dusk and regarded the picking season as a holiday with pay and the reunion of old friends. There was no machinery to spoil the talking and the family exchanged their cramped town life for a month's rural freedom. Unfortunately there was local hostility towards them in some areas and even goods in shops were removed from the customer's side of the counter and taken to the stores and cellars for safety. Gypsies and other travellers were attracted into hop-picking areas and it was common until the 1950s to see 'No dogs, gypsies or hoppers' signs displayed in many public houses. Other landlords would only serve hoppers outside through an open window. The hoppers seemed to accept the situation cheerfully, much as they had earlier accepted squalid accommodation. There was many a wrangle between pickers and the measurers, often because of leaves in

Bins were measured out at least twice a day. The measurer would arrive with his bushel measure accompanied by the booker, often a local schoolmistress, who carefully recorded the number of bushels picked. The measurer's assistant or a binman would always be on hand to hold open the poke to receive the freshly picked hops.

the bin, and although sometimes locals and Londoners worked happily side by side, many hop growers felt it prudent to segregate them into different parts of the garden.

The first English hop-picking machine was made at Suckley in Worcestershire in 1934 but machines were little used until 1948 and were first generally accepted in the west midlands. Some farmers resisted mechanisation because machine-picked hops were more difficult to dry as they settled unevenly in the kiln. But in the last days of hand picking, pay was about 2s 6d per bushel whereas hops picked by machine worked out at about 4d a bushel, thus economically justifying the installation of expensive machinery. Now there is hardly a farm left where picking is manual.

Today a tractor moves up and down the alleys and an operative on an elevated platform rapidly cuts down the bines which drop into a trailer. A more recent development is the automatic bine puller and loader.

ABOVE: *Hop picking also attracted gypsies and other travellers and it was a common sight even in the late 1950s to see a 'No dogs, gypsies or hoppers' sign displayed at many public houses, which were very crowded during the hop harvest.*

BELOW: *Hovering up. The picker thrusts his arms into the basket, lifting the hops up so they would lie as lightly as possible.*

ABOVE: *In east Kent hops were almost invariably picked into five-bushel tally baskets. Note the hop grower's initials painted on the side of the baskets and the cockscomb binman's hook.*
RIGHT: *Hop picking today. The bines are slashed with a knife about 2 feet (600 mm) above ground level and the strings are chopped from the wires from a crow's nest position. The bines then drop into a trailer below.*
BELOW: *An alley bodge or hop dolly, a miniature wagon used for transporting manure down the alleys. It could be pulled by a horse in either direction.*

ABOVE: *A load of freshly picked hops is about to be taken to the oast for drying. Oxen were still used in the Weald until the early part of the twentieth century.*

BELOW: *A load of cut bines on its way to the picking plant from a modern hop garden.*

Hop tokens of sixty, twelve and one bushel denomination.

Additional labour is still required for the hop harvest, for tractor work, to operate the machinery and for hand sorting to remove any pieces of stem or leaf. As with hand picking, the presence of too many hop leaves can cause severe problems when the hops are dried. Many of the hop picking machines supply 350 eight-bushel pokes in one day.

Hop tokens
Traditionally hop picking was piecework, paid according to the number of bushels picked, and metal discs known as tokens were used in place of ordinary money as far back as the end of the seventeenth century and were in use for at least 250 years. The grower would guarantee that the money they represented would be paid in full at the end of the harvest at the agreed rate. Originally tokens were only given for less than five bushels, but later they came into more general use and tokens for various numbers of bushels were made (rarely for money values) and were certainly more convenient and less cumbersome than the wooden tally sticks. Tokens were also known as 'checks' or 'medals'. They were accepted by local shopkeepers, innkeepers and the camp hawkers, who knew that the tokens would be honourably redeemed by the hop grower at the agreed rate at the end of the season.

Some of the earliest tokens were cast in pure lead and marked plainly with one initial of the grower's name. Later, to avoid confusion, two initials were used, often surrounded by dots. Tokens were produced in many shapes and sizes, some specially cast, others simply punched out and struck on one side only by the local builder, the

Sunday morning with the hop pickers. The camp pedlars accepted hop tokens in lieu of money for they knew that they would be redeemed without difficulty for cash at the end of the hop-picking season.

blacksmith, or the farm hands. The denominations varied greatly. Sets of 1, 6, 12, 30, 60 and even 120 bushels were among favourite denominations, but there was no general rule of value, shape, design or materials: tokens have been found made of lead, pewter, tin-coated iron, bronze, brass, copper, zinc and old bell metal.

The tally stick

The tally stick was a simple and infallible method of recording the number of bushels any particular worker had picked. The method was based on the old exchequer tally and comes from the French word *tailler* (to cut). The exchequer tally was introduced into England at the time of the Norman conquest, the old exchequer tallies being made of well seasoned sticks of hazel or willow. Tally sticks were in use in hop gardens possibly from as early as the seventeenth century and were certainly still in use in Kent in 1938.

The hop tally stick consisted of a length of wood 10 to 12 inches (250 to 300 mm) long rather like a ruler. The length was cut down the middle, longitudinally, parallel with its wider surface, for a little over

The tallyman at Newton Dilwyn, 1903.

Tally sticks.

three-quarters of its length, and one side was sawn off. The longest strip with the thicker unsawn end was retained by the tallyman and the smaller piece given to the picker. When the tallyman came round with the measurer, the two pieces were laid together and a notch scored across both parts, with a triangular file, for every five bushels picked. The picker kept the shorter, counter tally. The longer parts of the tallies, which were pierced at the top, were threaded on a cord which the tallyman hung on his belt or around his person. There could be no dispute over the quantity picked, for when put together the tallies would have to match exactly.

4. Drying the hops

The picturesque hop kilns of hop-growing districts, especially in the south, are well loved features of the landscape. In the south kilns are known as oasts and in the west midlands as kells. Today the word usually refers to the whole of the building — the receiving barn, the drying and cooling floor and the kiln itself. Every hop farm had at least one oast and disused oasts still remain in areas where hop growing has long since ceased, often converted into homes. Those oasts still used for hop drying are carefully maintained for use for just one month of the year.

Reynolde Scot described 'an oste as they dry their hoppes upon at Poppering', comprising a building 18 to 20 feet (5.5 to 6.1 m) long and 8 feet (2.4 m) wide with three rooms. The middle room contained a brick canopied furnace 13 inches (329 mm) wide and 6 or 7 feet (1.8 to 2.1 m) long and about 2 feet 6 inches (760 mm) high. The drying floor was about 5 feet (1.5 m) above the lower floor and composed of 1 inch (25 mm) square wooden laths laid a $\frac{1}{4}$ inch (7 mm) apart. Green hops were delivered into the first room, then carefully spread out on the drying floor in a layer about 18 inches (460 mm) deep. When the hops were dried they were pushed off the drying floor with a rake, into the third room where they were left to cool. Openings were provided between the upper part of the kiln and each of the end rooms. Modern oasts retain this basic layout.

Scot commented disapprovingly that 'Some lay their hops in the sun to dry and this taketh away the state of the hops, contrary to the purpose of drying which is very prejudicial to the brewer.' There are no surviving examples of Elizabethan oasts. Possibly the oldest example similar to Scot's description and probably the smallest remaining oast is near Cranbrook in Kent. It is constructed on an oak frame of reclaimed materials and appears to date from around 1750.

In Scot's day wood was used for hop drying; then charcoal was introduced and remained the standard fuel until the 1930s. There were plenty of charcoal burners in the hop-growing districts and a charcoal-burning industry still survives in Sussex and Hampshire. A hundred sacks of charcoal were needed to dry a ton of hops.

During the seventeenth century drying methods were varied and often barns were adapted to provide a drying floor with a fireplace installed below. Most purpose-built oasts up to the early eighteenth century were like barns with one or more kilns in the middle. The treatise *On the Planting and Management of Hops,* published by the Dublin Society in 1733, describes such an oast, and a brief

Sussex oasts. The small single-storey building attached to the stowage was probably the charcoal store.

specification mentioned in a lease of 1714 gives the size of a new oast as 40 feet (12 m) long and 17 feet (5 m) wide.

In 1739 Samuel Trowel describes an iron furnace with a closed grate and a vent, thus giving heat but keeping the smoke away from the hops. J. Mills in his *New System of Practical Husbandry* (1763) writes that 8 or 10 acres (3 to 4 ha) of hops require a building about 50 feet (15 m) long and 15 feet (4.5 m) wide, with chambers at each end for receiving and cooling respectively. As the hop acreage increased farmers endeavoured to dry a deeper bed of hops at one drying, so the flow of heated air had to be increased. Existing buildings were converted and new ones built which incorporated a pyramid-shaped ceiling, sometimes within the existing roof, gathering into an opening just above the ridge line. This acted as a flue to assist the draught. At the end of the eighteenth century the familiar protective wooden cowl, pivoted to turn away from the wind, was evolved. Traditionally the wooden cowls were made by the wheelwright and later examples were about 9 feet (2.7 m) in height with some as high as 14 feet (4.3 m): a disused cowl was once used as a bus shelter in the village of Charing in Kent. Often the vanes were embellished with finials, a favourite in Kent being the rampant horse.

The laths on the drying floor were spaced out to allow more air

ABOVE: *This is possibly the oldest example of an oast derived from Reynolde Scot's design and is probably the smallest remaining oast; it is near Cranbrook, Kent.*
BELOW: *A Kentish barn oast. A pyramidal ceiling was formed within the roof structure, and only the cowl covering the opening just above the ridge line reveals that the building is an oast.*

ABOVE: *Traditionally it has been the job of the village wheelwright to make cowls, and ideally they are constructed from well seasoned hardwood. Here the wheelwright is ascending with the cowl to prevent damage to both the kiln and the cowl.*

BELOW: *Three early nineteenth-century ragstone oasts in a remote Kentish valley near Sevenoaks, now owned by the National Trust. The octagonal cowls are extremely rare.*

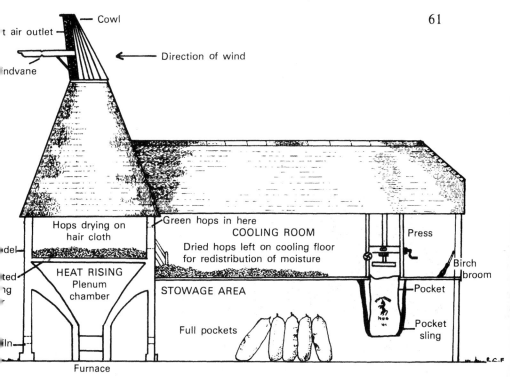

A section through a Victorian oast.

through and the hops were placed on a horse hair cloth which resisted the sulphur fumes, an arrangement still in use today. Sometimes perforated tiles and wire were used. Kilns were up to 16 feet (5 m) square, and multiple oasts beside single stowage were built – often as many as eight or ten kilns in a row, fired by charcoal, coke and charred pit coal.

The oast with the round kiln was introduced at the beginning of the nineteenth century, especially in the south-east. An early example, dated 1815, can be seen at the Wye College Agricultural Museum at Court Lodge Farm, Brook, Kent. The design was improved in 1835 by John Read from Horsmonden in west Kent, whose monument in Horsmonden church (which is surrounded by hop gardens) records that he was the inventor of the stomach pump 'and many other implements for the benefit or relief of humanity'. In the last quarter of the nineteenth century, two 18 foot (5.5 m) diameter kilns with cooling room cost about £500 to build. With the improved design, twice as many hops could be dried with better results because of the increased distance between the hop fire and the kiln. Often multiple

LEFT: *This oast is immediately identifiable as a west midland example because of the pointed 'witch's hat' type cowl. Oasts in the south-east almost invariably have a flat top.*

BELOW: *This oast near Hereford retains one traditional west midland style cowl but the others have been replaced with louvred vents.*

Frequently square oasts and roundels built at different dates are found in the same oast complex.

fireplaces were provided for better heat distribution.

After about 1900 it was discovered that square kilns distributed heat just as evenly and as these were easier to construct, most oasts built after 1900 were 18 to 20 feet (55 to 60 m) square. Electric fans became general by 1930 and louvred openings often replaced revolving cowls. Oil was first used in 1933 and today there are very few oasts that have not been converted to oil heating. By 1937 the cost of erecting an oast house had risen to £10,000.

Freshly picked hops contain about eighty per cent moisture, which must be reduced by drying to about six per cent. During storage it will probably rise again to about 10 per cent. It was known as far back as Reynolde Scot's time that the main object of drying is not to let the hops dry out completely but to ensure that the moisture is evenly distributed. This calls for the utmost care and at harvest time the dryer in the oast is the most important man on the farm. He has often worked with hops all his life and during those few weeks in September he must be on hand twenty-four hours a day. He also decides when picking should stop. He has a camp bed in the upper drying floor adjacent to the kiln so that he can tend his hops and regulate the heating during the night. Today good dryers are few and training new dryers is difficult as there are only three or four weeks in the year to gain practical experience of how to feel, smell and recognise the exact degree of dryness required. Often hop dryers are brought out of retirement each year to help.

ABOVE: *One of the most impressive arrays of oasts, at Whitbread's hop farm at Beltring near Paddock Wood, Kent, in the centre of the south-eastern hop-growing district.*

LEFT: *When a round kiln was slated, each slate had to be tapered individually. Clay tiles were manufactured specially for kiln roofs and were made slightly wedge-shaped to allow for the conical formation.*

In traditional kilns the hops are spread about 2 feet (600m) deep and until 1980 sulphur was introduced at the start of the drying process, which turned the green cones yellow and lessened discolouration, giving the sample a bright and uniform appearance. However, in 1980 dryers were instructed by the Hops Marketing Board, at the request of the Brewers Society, not to sulphur hops any more and most are relieved that this operation has ceased.

Until quite recently the dryer had no technical instruments: by using his nose and hands he would judge when the hops were sufficiently dry. Now much lighter in weight, they are removed from the kiln on to the floor of the cooling loft to cool and stabilise the moisture content. Then large canvas or hessian backed shovels known as *scuppets* are used to scoop the hops into the pocket, which is suspended under a hole in the cooling loft floor beneath the press. To prevent deterioration of the hops it is essential that they are compressed firmly in the pocket. As the pocket fills up the hops are rammed down tightly by turning the winch of the press. This process is repeated several times until the pocket is tight and round, filled almost to bursting point and weighing about 1½ hundredweight (75 kg). The neck is sewn up by hand, leaving two ears by which the pocket can be handled. The pocket sling, which steadied the pocket whilst pressing, is released and the pocket is dropped down to the floor below and taken to the hop warehouses for storage and sampling.

The hop press is believed to have been invented in about 1850 by Mr Ellis from Barming near Maidstone, who, in 1835, had 500 acres

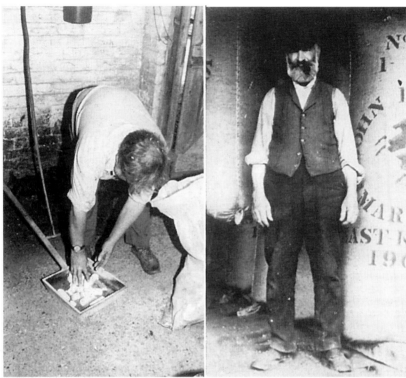

ABOVE LEFT: *Placing rolled sulphur in the brimstone pan. It was the custom in England for many years (except in the Farnham area) to introduce sulphur into the furnace at the commencement of drying. It gave the hops a bright yellow appearance, which was what the hop merchant required. The use of sulphur was challenged many times over the years but sulphuring did not cease until 1980.*

ABOVE RIGHT: *The hop dryer is the most important person during the hop harvest for he must ensure that the hops dry evenly, the fires are at the correct temperature, and the hops are cured to precisely the right texture. Frequently dryers spend twenty-four hours a day in the oast, taking only catnaps on a makeshift bed.*

(200 ha) of hops when no other single grower had more than 150 acres (60 ha). Today most presses are power-operated although there are still several which are hand-operated and within living memory hops have been pressed into the pocket by the old method of treading. The man who trod the hops into the pockets was known as the *bagster* and wore special shoes and a straw hat. He stepped into the pocket from a small ladder and whilst his companions fed in the hops with the scuppet, the bagster trod them firm and continued to do so

RIGHT: *Stoking the furnaces in the open traditional kilns demanded some practice to obtain the various temperatures required. One method of regulating the fire was by adjusting the sliding door (on the left of the picture) opposite the fire.*

BELOW: *Loading the kiln. To ensure uniformity of loading and to avoid any compression, the hops are delicately 'hovered up'. The dryer here is using a blunt-tined fork.*

ABOVE: *After drying the hops are very fragile and they are laid out on the floor of the cooling loft to equalise the moisture content in the cones, which become tougher and more suitable for pressing.*

BELOW: *Scuppeting the hops towards the press.*

Pressing the hops. To preserve the hops they are packed so tightly that they eventually have to be cut out of the pocket for use.

until his head appeared above floor level and the pocket was full. Since 1979 there has been increasing use of the hydraulic hop baler, which produces a rectangular bale of hops of similar weight to the traditional pocket.

A visit to a working oast during September is a pleasing experience. In the old coal-fired oasts the sulphur made the flames burn bright blue in the furnace below. In the cooling room above, all is clean and orderly; a large heap of dried hops is spread out on the floor, and the fillers are steadily scuppeting the light hops into the pocket. The floor is continually swept with a traditional broom and occasionally the dryer wades through his cooking hops, always watching and feeling, and throughout the oast there is the smoky bitter-sweet aroma of drying hops.

Examination and valuation

The final value of the crop is assessed by the condition of the hop samples and the skills of a good hop sampler were greatly respected. After hop drying the samplers often worked all night to sample as much as possible in time for the market: one Wealden hop sampler, practising in the early years of the twentieth century, cut open and sewed up sixty pockets in one hour for a wager.

Until the last war hops would often be sampled in the warehouse of the hop factor immediately after sale by sample to the merchant. The

LEFT: *A scuppet, Hampshire pattern.*

Scuppets are still made by hand in a country workshop.

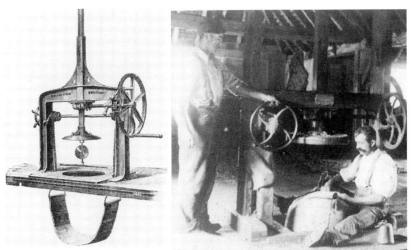

ABOVE LEFT: *The 'Improved Pattern Press', costing £13 13s when first introduced.*
ABOVE RIGHT: *When the pocket is full and pressed almost to bursting point it is sewn up by hand, using a strong coping twine and a long curved coping needle. This method is still used today.*

merchant's examiner would also be present with his hop sampling knife and tools. The pocket was weighed and once it had been accepted by the examiner there was no redress if a pocket was subsequently found to be faulty. Thus the examiner's job was one of immense responsibility. This function is now undertaken by a representative of the grower's producer group who will also have a person authorised by the Ministry of Agriculture, Fisheries and Food whose function is to certify that the hops meet the EC standards. However, the same sampling techniques and skills are still used in determining the condition of the hops.

Often when the hops are first received from farms they are still settling in the pocket and taking up moisture, so they are not sampled immediately. As the samples are still graded primarily by visual appearance, considerable effort is made to ensure that each sample is prepared in precisely the same way. The samples are now taken with the pockets lying down, but in the past samples were often taken from the pocket standing up. To draw the sample, the side of the pocket is opened and held back with a pair of skewers or *spreaders* and two parallel longitudinal cuts are made with a long sharp knife. The sampler thrusts a large pair of pincer-like tongs, known as *clams*, into the pocket between the cuts and draws out the sample, which is trimmed into a neat rectangular block and quickly wrapped in strong paper and numbered. The space left in the pocket is filled by hops kept at hand especially for the purpose and the pocket is sewn up again and reweighed.

ABOVE: *The pocket is suspended beneath the press and is supported by webbing or a pocket sling.*
BELOW: *In earlier days all growers tried to be the first to get their hops to market, for the earliest hops always fetched a good price. Usually the pockets were arranged on the wagon with the blank unstencilled side on view, so other hop growers would not see the numbering of the pockets, from which they would have been able to tell how successful that grower's harvest had been. Many growers would even start numbering their pockets at 11, 51 or even 101 for the same reason.*

An experienced sampler can tell from the feel of the cutting action whether the hops are under-cured or perished. An examiner will pick a handful of hops, examine the colour and rub them in one hand with his thumb. From this simple test by sight, feel and aroma, he can judge the quality of the sample. In recent years the electric moisture tester has been introduced as an aid to examination. The two prongs are inserted in the pocket and an electric current passes between them. The conductivity of the hops will vary according to the moisture content, which is read from a calibrated scale.

The samples are graded by each marketing cooperative. All 'high alpha' hops are analysed in a recognised laboratory for alpha acid content. The price is then adjusted up or down according to the individual analysis of the sample relative to the norm for that particular variety. Hop growers are entitled to have the grading decision reviewed, and finally there is the right for either party to appeal to an independent panel.

BELOW LEFT: *Sampling. Two parallel cuts are carefully made with the sampling knife.*
BELOW RIGHT: *Old sampling tools. The double-bladed knife is no longer used today as most samplers prefer to use a single-bladed knife, always kept razor sharp.*

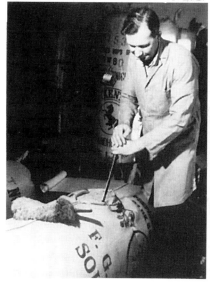
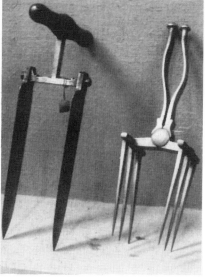

5. Other uses for hops

Hop shoots were a delicacy from Roman times and buttered hop tops and hop sauce were frequently prepared, certainly in Kent, in comparatively recent years.

Old recipe for buttered hop tops
Before May gather the young hop shoots about four leaves down and soak them in a bowl of cold water with a handful of salt. Drain and plunge into enough white water to cover the hops easily and boil rapidly. When they are tender, drain the water off immediately and chop them up in the hot pan with pepper, salt and a lump of butter. Serve hot.

Hop sauce
Prepare as above and chop into a previously prepared butter sauce. Serve with chicken, fish or mutton.

Hops were believed to have protective and healing properties. Sprays of hops were carried before assize judges on prison visits to protect them from the foul air. The leaves were made into a medicine to relieve certain skin diseases and a bag of hops soaked in boiling water was used as a poultice to relieve stiffness and pain. Cows were often fed hop leaves to increase their milk yield.

Hops are also supposed to have soporific qualities. It was said that some people entering an oast during the hop drying fell asleep and one method of inducing sleep was to inhale the steam from an infusion of hops. The best known and most widely used remedy is to lie on a pillow stuffed with hops: George III tried it, on the advice of Henry Addington, and it worked. Hop pillows are still made today.

Spent hops — the part that remains after brewing — are used as plant food and hop bines have been used to produce a coarse woven material and to provide *lewing*, a windbreak for the hop gardens.

6. Glossary

Ale: the English term for unhopped malt beverages, as distinct from beer, which was brewed with hops.

Alley: the space between two rows of hops.

Alley bodge: a miniature wagon used for carting manure along the alleys.

Alpha acid: a bittering substance contained in hops.

Anchor wire: the guy wire attached to the straining pole in modern wirework.

Bagging machine: the machine for tightly pressing dried hops into the pocket.

Bagster: the workman who stamped or trod dried hops into the pocket or bag before the hop press was introduced.

Banding in: in some systems of hop stringing, the strings are drawn together about 5 feet (1.5 m) from the ground, keeping the alleys clear and giving greater rigidity to the strings.

Barracks: a term for hoppers' accommodation during harvest.

Bat: a general term for the pole or even simply a length of wood.

Beta acid: a bittering substance contained in hops.

Bin: a portable collapsible wooden framework with a loose open-topped hessian bag, used for containing the freshly picked hops, usually holding about 20 bushels.

Bine: the stem of the hop.

Binman: the overseer in charge of the bins.

Binman's hook: the bine hook or long-poled hop knife used for un-hitching the strings from the top of the pole.

Blight: infestation of the hop by aphis.

Booker: the man who recorded the amount of hops picked by each picker.

Bushel: 8 gallons, dry measure (about 36 litres).

Butcher system: a method of hop training on wirework.

Canterbury hoe: three-tined long-handled hoe used in the hop garden.

Check: hop token (qv).

Checker: booker (qv).

Cockle: an enclosed iron stove used to heat a kiln.

Coir: durable string made from coconut fibres.

Cooling: the period between drying and pressing when hops are left on the cooling room floor for the remainder of the moisture content to be redistributed evenly.

Coping: sewing up the opening of the hop pocket.

Cowl: the hood mounted on the top of an oast, pivoted so that its opening always faces away from the wind, preventing down-draughts.

Crib: bin (qv).

Crow's nest: platform on a tractor-pulled trailer adjusted to a suitable height for cutting hop bines and enabling elevated work to be carried out.

Curing: drying the hops by reducing the moisture.

Downy mildew: a fungus disease.

Dressing knife: the knife used to cut the roots around the hill.

Dryer: skilled worker in charge of the oast and responsible for drying the hops.

Drift: formation of an orderly line to picking bins in the hop garden.

Dry hopping: putting a small quantity of hops or hop pellets into a full cask of beer before it leaves the brewery.

Dusting: applying sulphur fungicide or insecticide in the form of powder to the growing hops.

Feathering: the condition of the hop cone when it is partially dried and the bracts tend to stand out giving the cone a feathery appearance.

Footshoe money: an old tradition where visitors to the hop garden would have their shoes rubbed with a handful of hops. A contribution was then made to the hoppers' feast at the end of the hopping season.

Foxy: description given to hops that had poor colour before drying and would not give a good sample.

Garden: hop plantation.

Greenbag: poke (qv).

Green hops: freshly picked, undried hops.

Green stages: an open slatted floor adjoining the drying floor where the pokes are temporarily stored before drying.

Hair: woven horsehair placed on the drying floor of the oast supporting the hops but allowing the heat from the furnace to circulate.

Hill: single hop plant or group of hop shoots growing together.

Hog: an iron baffle plate fixed over an open furnace in an oast kiln protecting the hops from a stray spark.

Hop dog: 1, a simple implement with a wooden handle on to which notched iron jaws were fixed, used for prising up hop poles. 2, the caterpillar stage of the pale tussock moth, found in the hop gardens from August to October.

Hop extract: a soluble bittering extract of hops concentrated into a syrup.

Hop fork: a wooden, blunt-tined fork used in the oast for ensuring that the hops lay evenly in the kiln.

Hoppers' hut: a shelter or small building provided by the hop grower for the hop pickers.

Hop pellets: compressed hop powder.

Hop pole: the pole up which the bines are trained.

Hop token: a metal disc used instead of money to guarantee payment to the pickers.

Horse: hog (qv).

Horse pole: term used in the Weald in the days of the pole system. At the start of picking a stripped pole, known as a horse pole, was placed across the two upright members at each end of the bin. The hop poles were then placed on it for easier picking.

Hovering up: loosening up the hops on the drying floor of the kiln to avoid compaction or in the bin just before the measurer came round.

Huffkin: a special cake made in Kent for the hoppers' feast at the end of the hop-picking season.

Keele: furnace in an oast.

Kell: oast (qv).

Knippers: huge iron or wooden toothed pincers used for drawing out hop poles.

Lanthorn: eighteenth-century word for a complete furnace unit.

Legging: stripping the leaves of the lower bines:

Lewing: a windbreak of hedges, coir yarn, plastic, a tarpaulin or even old hop poles around the hop garden.

Lifter cloth: a cloth hung on hooks on the wall of the drying kiln covering sections of the hair. Each section could be picked up to lift the hops out of the kiln on to the drying floor.

Lobber: a weight to which coir yarn could be attached and thrown over the poles to save a lot of stilt or ladder work. The lobber was frequently used in early string systems.

Lupulin: resin glands found at the base of the cone bracts.

Measurer: the man who measured the number of bushels of hops transferred from the bin to the poke.

Medal: hop token (qv).

Mudging: the effect of spoiling in hops which perhaps were not sufficiently cool to press well.

Nidget: a triangular shaped horse-drawn hoe used for tilling the alleys of hop gardens.

Oast: building used for the drying, stowage and pressing of hops.

Peeler: an iron bar with a swollen base reducing to a point, used to make the holes for hop poles.

Pipey bines: the quick-growing stalks of the hop set showing promise of much stem but little fruit. They are pinched out in May.

Pitcher: peeler (qv).

Plenum chamber: the part of the oast surrounding the furnace.

Pocket: sack into which the hops are pressed after drying.

Pocket sling: the webbing underneath the press which holds the pocket steady and gives additional support during pressing.

Poke: bag of loosely woven material, containing about 10 bushels, in which the green hops were carried to the oast after they had been measured out from the bins. Also known as a green sack and still commonly used to transport hops from the picking machine.

Pole work: the original system of growing hops where the hops are trained directly up the poles.

Press: bagging machine (qv).

Pressing: the compression of dried hops into the pocket.

Reeked: the term given by hop dryers to the condition of hops that have acquired a dull appearance.

Riffle board: a hop dresser would often tie a riffle board to his leg with a sharpening stone attached so that he could sharpen the hop dressing knives without having to stand up. Often the hop dressers were paid on piecework so time was important to them.

Roguing: the process of pinching out the pipey bines.

Roundel: a circular kiln of an oast.

Sample: rectangular section of hops cut out from the side of the pocket for inspection, valuation and sale purposes.

Sampling hook: the hook that holds back the cut seam in the hop pocket whilst the sample is removed.

Sampling knife: sharp double or single bladed knife used for making the cuts when taking a sample.

Saplier: hop cloth used in the Surrey gardens for gathering up the hops.

Screw peg: peg of galvanised wire with a pig's-tail twist, screwed into the ground near the hill where the string can run through the twists.

Scuppet: wooden shovel with a canvas or hessian back used for moving the hops around the oast.

Set: 1, hop plant grown for transplanting (often spelt 'sett'). 2, in pole work, a block of one hundred hills.

Slack pocket: a pocket of hops holding too much moisture, which will deteriorate rapidly.

Spreader: sampling hook (qv).

Steddle: lanthorn (qv).

Spud: a heavy three or four tongued fork with flat tines used for digging the alleys.

Stilts: pair of wooden or metal poles with foot rests and a leather belt, used for elevated working.

Stock: hill (qv).

Stringer: one who strings up the hop plantation.

Stringing rod: length of wood (now almost invariably cane) long enough to reach comfortably the top wires, often with an angled piece of piping at the end, used for continuous stringing of wire work.

Sulphuring: 1, burning rolled sulphur in the furnace at the early stage of drying. 2, applying powdered sulphur to the bines as a fungicide.

Surplices: Hampshire term for picking cloths.

Tally: 1, hop tally—the split wooden slat for recording the number of bushels of hops picked. 2, the tally—the agreed rate of payment; traditionally the number of bushels picked for one shilling.

Tommy hoe: short-handled hoe used when the hops are dressed.

Training: the process of gently twirling the bines up the strings.

Twiddling: Kentish expression for training.

Umbrella system: method of stringing.

Verticillium wilt: a fungus disease.

Wirework: system of poles supporting wires to which strings are attached for the hop plants to climb.

Worcester system: method of stringing.

Yard: west midlands expression for hop plantation.

ACKNOWLEDGEMENTS

Illustrations are acknowledged as follows: Botting Collection, Ashford Heritage Centre, page 42 (bottom); Brewers Society, pages 22, 72 (top); R. Chalk, page 43 (top left); Dormington Hop Farm, page 37; Egerton School, page 48; Hereford Public Library, pages 24, 33 (bottom), 55; P. C. Hilton, page 60 (top); *Kent Messenger*, page 68 (bottom); Stuart and Dorothy Murray, pages 40 (top), 49 (bottom), 66 (right); Robert Swift, page 45 (top) and front cover; Weaver Brothers Library, pages 32, 34 (bottom right), 71 (right); Whitbread and Company, pages 46 (bottom) and 52 (bottom). Remaining illustrations are by the author or from his collection.

The publishers and author acknowledge their indebtedness to the following companies, organisations and individuals who in various ways have assisted in the preparation of this volume: Allied Breweries; Ashford Heritage Centre; The Brewers Society; Christine Swift Antiquarian Bookshop, Maidstone; The County Council of Hereford and Worcester; Cranbrook Galleries; Curtis Museum, Alton; Dormington Hop Farms Limited; Hop Products; Hops Marketing Board; *Kent Messenger;* Maidstone Museum; Tenterden Museum; Twymans; Whitbread and Company; Wolseley Place Studios, Ashford; Wye College; Alf and Trevor Austen; Ernie Beale; Ken Carley; Fred Chapman; R. John Cyster, OBE; P. Day; Arthur Farley; R. Farrar; H. Hilder; Alf Horton; F. Huntingford; N. L. Kendon; Arnold Kingsnorth; Charlie Leonard; David Millar; John Paine, BSc; Robert Swift; George Weaver; as well as to many other hop growers, dryers, pickers, stringers, hop garden erectors and others involved in the hop industry who over the years have given specialist knowledge including I. S. Wordsworth of English Hops Limited.

The publishers and the author are grateful to Bob Farrar, formerly of the Department of Hop Research, Wye College, and to John Lander, Hop Consultant, for their advice and assistance.

Index